ANIMAL TRACKS

of

MISSISSIPPI

&

LOUISIANA

I0112448

Tamara Eder

with contributions from Ian Sheldon

LONE
PINE

THE PUBLISHER: LONE PINE PUBLISHING

1901 Raymond Avenue SW, Suite C	10145-81 Avenue
Renton, WA 98055	Edmonton, AB T6E 1W9
USA	Canada

Website: http://www.lonepinepublishing.com

National Library of Canada Cataloguing in Publication Data

Eder, Tamara, (date)
  Animal tracks of Mississippi and Louisiana

  Includes bibliographical references and index.
  ISBN 1-55105-315-2

  1. Animal tracks—Mississippi—Identification. 2. Animal tracks—Louisiana—
Identification. I. Sheldon, Ian, (date) II. Title.
QL768.E36 2001   591.47′9   C2001-910206-2

  Editorial Director: Nancy Foulds
  Editor: Volker Bodegom
  Proofreaders: Randy Williams, Lee Craig, Genevieve Boyer
  Production Coordinator: Jen Fafard
  Design, Layout and Production: Volker Bodegom, Monica Triska
  Cover Design: Elliot Engley
  Cartography: Volker Bodegom
  Technical Contributor: Mark Elbroch
  Animal Illustrations: by Gary Ross, except for those by Ewa Pluciennik
    (p. 107) and Ted Nordhagen (pp. 95, 111).
  Track Illustrations: Ian Sheldon
  Cover Illustration: Raccoon by Gary Ross
  Scanning: Elite Lithographers Ltd.

We acknowledge the financial support of the Government of Canada
through the Book Publishing Industry Development Program (BPIDP)
for our publishing activities.

PC: P4

CONTENTS

INTRODUCTION

If you have ever spent time with an experienced tracker, or perhaps a veteran hunter, then you know just how much there is to learn about the subject of tracking and just how exciting the challenge of tracking animals can be. Maybe you think that tracking is no fun, because all you get to see is the animal's prints. What about the animal itself—is that not much more exciting? Well, for most of us who don't spend a great deal of time in the beautiful wilderness of Mississippi and Louisiana, the chances of seeing the elusive Red Fox or the fun-loving River Otter are slim. The closest that we may ever get to some animals will be through their tracks, and they can inspire a very intimate experience. Remember, you are following in the footsteps of the unseen—animals that are in pursuit of prey, or perhaps being pursued as prey.

This book offers an introduction to the complex world of tracking animals. Sometimes tracking is easy. At other times it is an incredible challenge that leaves you wondering just what animal made those unusual tracks. Take this book into the field with you, and it can provide some help with the first steps to identification. Animals tracks and trails are this book's focus; you will learn to recognize subtle differences for both. There are, of course, many additional signs to consider, such as scat and food caches, all of which help you to understand the animal that you are tracking.

It takes many years to become an expert tracker. Tracking is one of those skills that grows with you as you acquire new knowledge in new situations. Most importantly, you will have an intimate experience with nature. You will learn the secrets of the seldom seen. The more you discover, the more you will want to know, and, by developing a good understanding of tracking, you will gain an excellent appreciation of the intricacies and delights of our marvelous natural world.

How to Use This Book

Most importantly, take this book into the field with you! Relying on your memory is not an adequate way to identify tracks. Track identification has to be done in the field, or with detailed sketches and notes that you can take home. Much of the process of identification involves circumstantial evidence, so you will have much more success when standing beside the track.

This book is laid out in an easy-to-use format. Beginning on p. 124, there is a quick reference appendix to the tracks of all the animals illustrated in the book. This appendix is a fast way to familiarize yourself with certain tracks, and it guides you to the more informative descriptions of each animal.

Each description is illustrated with the appropriate footprints and the track patterns that it usually leaves. Although these illustrations are not exhaustive, they do show the tracks or groups of prints that you will most likely see. You will find a list of dimensions for the tracks, giving the general range, but there will always be extremes, just as there are with people who have unusually small or large feet. Under the category 'Size' (of animal), the 'greater-than' sign (>) is used when the size difference between the sexes is pronounced.

If you think that you may have identified a track, check the 'Similar Species' section. This section is designed to help you confirm your conclusions by pointing out other animals that leave similar tracks and by showing you ways to distinguish among them.

As you read this book, you will notice an abundance of words such as 'often,' 'mostly' and 'usually.' Unfortunately, tracking will never be an exact science; we cannot expect animals to conform to our expectations, so be prepared for the unpredictable.

Tips on Tracking

As you flip through this guide, you will notice clear, well-formed prints. Do not be deceived! It is a rare track that will ever show so clearly. For a good, clear print, the perfect conditions are slightly wet, shallow snow that is not melting or slightly soft mud that is not actually wet. These conditions can be rare—most often you will be dealing with incomplete or faint prints where you cannot even really be sure of the number of toes.

Should you find yourself looking at a clear print, then the job of identification is much easier. There are a number of key features to look for: measure the length and width of the print, count the number of toes, check for claw marks and note how far away they are from the body of the print, and look for a heel mark. Keep in mind more subtle features, such as the spacing between

TENNESSEE

ARKANSAS

Tallahatchie R.

TUPELO

MISSISSIPPI

GREENVILLE

Yazoo R.

Big Black R.

Red R.

MONROE

SHREVEPORT

TALLULAH

VICKSBURG

MERIDIAN

JACKSON

LOUISIANA

Ouachita R.

NATCHITOCHES

Toledo Bend Res.

NATCHEZ

Pearl R.

HATTIESBURG

ALEXANDRIA

Mississippi R.

BATON ROUGE

ALABAMA

TEXAS

LAKE CHARLES

LAFAYETTE

Calcasieu Lake

NEW IBERIA

MORGAN CITY

Marsh Island

Atchafalaya Bay

HOUMA

Point au Fer Island

Terrebonne Bay

Lake Pontchartrain

GULF PORT

BILOXI

PASCA-GOULA

Chandeleur Sound

NEW ORLEANS

Breton Sound

Mississippi Delta

N

0   100 km

0   100 mi

Gulf   of   Mexico

7

the toes, whether or not they are parallel, and whether fur on the sole of the foot has made the print less clear.

When you are faced with the challenge of identifying an unclear print—or even if you think that you have made a successful identification from one print alone—look beyond the single footprint and search out others. Do not rely on the dimensions of one print alone, but collect measurements from several prints to get an average impression. Even the prints within one trail can show a lot of variation.

Try to determine which is the fore print and which is the hind, and remember that many animals are built very differently from humans, having larger forefeet than hind feet. Sometimes the prints will overlap, or they can be directly on top of one another in a direct register. For some animals, the fore and hind prints are pretty much the same.

Check out the pattern that the tracks make together in the trail and follow the trail for as many paces as is necessary for you to become familiar with the pattern. Patterns are very important and can be the distinguishing feature between different animals with otherwise similar tracks.

Follow the trail for some distance—it may give you some vital clues. For example, the trail may lead you to a tree, indicating that the animal is a climber, or it may lead down into a burrow. This part of tracking can be the most rewarding, because you are following the life of the animal as it hunts, runs, walks, jumps, feeds or tries to escape a predator.

Take into consideration the habitat. Sometimes habitat alone will allow you to distinguish very similar tracks—one species might be found on the riverbank, whereas another might be encountered just in dense forest.

Think about your geographical location, too, because some animals have a limited range. This consideration can rule out some species and help you with your identification.

Remember that every animal will at some point leave a print or trail that looks just like the print or trail of a completely different animal!

Finally, keep in mind that if you track quietly, you might catch up with the maker of the prints.

Nine-banded Armadillo

Terms & Measurements

Some of the terms used in tracking can be rather confusing, and they often depend on personal interpretation. For example, what comes to your mind if you see the word 'hopping'? Perhaps you see a person hopping about on one leg, or perhaps you see a rabbit hopping through the countryside. Clearly, one person's perception of motion can be very different from another's. Some useful terms are explained below, to clarify what is meant in this book and, where appropriate, how the measurements given fit in with each term.

The following terms are sometimes used loosely and interchangeably—for example, a rabbit might be described as 'a hopper' and a squirrel as 'a bounder,' yet both leave the same pattern of prints in the same sequence.

Ambling: Fast, rolling walking.

Bounding: A gait of four-legged animals in which the two hind feet land simultaneously, usually registering in front of the fore prints. It is common in rodents and the rabbit family. 'Hopping' or 'jumping' can often be substituted.

Gait: Describes how an animal is moving at some point in time. Different gaits result in different observable trail characteristics.

Galloping: A gait used by animals with four legs of even length, such as dogs, moving at high speed, hind feet registering in front of forefeet.

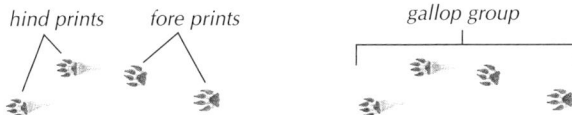

Hopping: Similar to bounding. With four-legged animals it is usually indicated by tight clusters of prints, fore prints set between and behind the hind prints. A bird hopping on two feet creates a series of paired tracks along its trail.

Loping: Like galloping, but slower, with each foot falling independently and leaving a trail pattern that consists of groups of tracks in the sequence fore-hind-fore-hind, usually roughly in a line.

Mustelids (weasel family) often use ***2×2 loping***, in which the hind feet register directly on the fore prints. The resulting pattern has angled, paired tracks.

Running: Like galloping, but applied generally to animals moving at high speed. Also used for two-legged animals.

Trotting: Faster than walking, slower than running. The diagonally opposite limbs move simultaneously; that is, the right forefoot with the left hind, then the left forefoot with the right hind. This gait is the natural one for canids (dog family), short-tailed shrews and voles.

hind print

fore print

Canids may use **side-trotting,** a fast trotting in which the hind end of the animal shifts to one side. The resulting track pattern has paired tracks, with all the fore prints on one side and all the hind prints on the other.

Walking: A slow gait in which each foot moves independently of the others, resulting in an alternating track pattern. This gait is common for felines (cat family) and deer, as well as wide-bodied animals, such as bears and porcupines. The term is also used for two-legged animals.

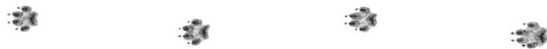

Other Tracking Terms:

Dewclaws: Two small, toe-like structures set above and behind the main foot of most hoofed animals.

Direct Register: The hind foot falls directly on the fore print.

double register   *direct register*

Double Register: The hind foot registers and overlaps the fore print only slightly or falls beside it, so that both prints can be seen at least in part.

Dragline: A line left in snow or mud by a foot or the tail dragging over the surface.

dragline

Gallop Group: A track pattern of four prints made at a gallop, usually with the hind feet registering in front of the forefeet (see **'galloping'** for illustration).

Height: Taken at the animal's shoulder.

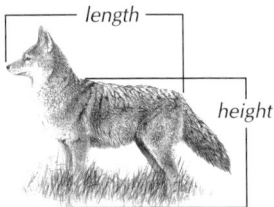
length

height

Length: The animal's body length from head to rump, not including the tail, unless otherwise indicated.

Metacarpal Pad: A small pad near the palm pad or between the palm pad and heel on the forefeet of bears and mustelids (members of the weasel family).

Print (also called '*track*'): Fore and hind prints are treated individually. Print dimensions given are 'length' (including claws—maximum values may represent occasional heel register for some animals) and 'width.' A group of prints made by each of the animal's feet makes up a track pattern.

Register: To leave a mark—said about a foot, claw or other part of an animal's body.

Retractable: Describes claws that can be pulled in to keep them sharp, as with felids (the cat family); these claws do not register in the prints. Foxes have semi-retractable claws.

Sitzmark: The mark left on the ground by an animal falling or jumping from a tree.

Straddle: The total width of the trail, all prints considered.

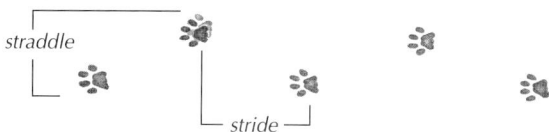

Stride: For consistency among different animals, the stride is taken as the distance from the center of one print (or print group) to the center of the next one. Some books may use the term 'pace.'

Track: Same as '*print*.'

Track Pattern: The pattern left after each foot registers once; a set of prints, such as a gallop group.

Trail: A series of track patterns; think of it as the path of the animal.

Red Fox

MAMMALS

River Otter

White-tailed Deer

Fore and Hind Prints
Length: 2–3.5 in (5–9 cm)
Width: 1.6–2.5 in (4–6.5 cm)

Straddle
5–10 in (13–25 cm)

Stride
Walking: 10–20 in (25–50 cm)
Galloping: 6–15 ft (1.8–4.5 m)

Size (buck>doe)
Height: 3–3.5 ft (90–110 cm)
Length: to 6.3 ft (1.9 m)

Weight
120–350 lb (55–160 kg)

walking   *gallop group*

WHITE-TAILED DEER

Odocoileus virginianus

The keen hearing of this deer guarantees that it knows about you before you know about it. Frequently, all that we see is its conspicuous white tail in the distance as it gallops away, earning this deer the nickname 'Flagtail.' This adaptable deer may be found throughout the region in small groups at the edges of forests and in brushlands. The White-tailed Deer can be common around ranches and residential areas.

This deer's prints are heart-shaped and pointed. Its alternating walking track pattern shows the hind prints direct registered or double registered on the fore prints. In mud, or when a deer gallops on soft surfaces, the dewclaws register. This flighty deer gallops in the usual style, leaving hind prints ahead of fore prints, with toes spread wide for steadier, safer footing.

Similar Species: The Feral Pig (p. 20) makes similar prints, but with a shorter stride, a wider gap between the toes and prominent dewclaws angled to the side.

Feral Pig

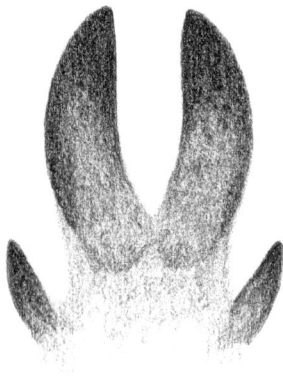

**Fore and Hind Prints
(with dewclaws)**
Length: 2.5–3 in (6.5–7.5 cm)
Width: 2.3 in (5.8 cm)
Straddle
5–6 in (13–15 cm)
Stride
Trotting: 16–20 in (40–50 cm)
Size (male>female)
Height: 3 ft (90 cm)
Length: 4.3–6 ft (1.3–1.8 m)
Weight
77–440 lb (35–200 kg)

trotting

FERAL PIG (Wild Pig, Wild Boar)

Sus scrofa

Descended from introduced European animals, the Feral Pig interbred with escaped Domestic Pigs (different in appearance but the same species). This sturdy beast of the dense undergrowth can be found in small, scattered populations throughout the region.

The Feral Pig's print shows two prominent, widely spaced toe marks, and usually, except on firm surfaces, a clear, pointed dewclaw mark off to each side. The hind print is slightly smaller than the fore print. Feral Pigs are keen foragers, so their tracks can often be numerous, especially when they travel in a group. The Feral Pig usually trots—a typical alternating track pattern shows a double register of hind print on fore print. Other signs are wallows and diggings.

Similar Species: White-tailed Deer (p. 18) prints can be similar, but they are slightly larger with a longer stride and a narrower gap between the toes; a deer's dewclaws may lightly register, but at the rear of the print, not the sides. Although the Domestic Pig is the same species, its tracks have a wider straddle and are less neat, often forming two separate lines.

21

Horse

Fore Print
(hind print is slightly smaller)
Length: 4.5–6 in (11–15 cm)
Width: 4.5–5.5 in (11–14 cm)
Straddle
2–7.5 in (5–19 cm)
Stride
Walking: 17–28 in (43–70 cm)
Size
Height: to 6 ft (1.8 m)
Weight
to 1500 lb (680 kg)

walking

HORSE
Equus caballus

Wilderness adventures on horseback are a popular activity, so you can expect horse tracks to show up almost anywhere.

Unlike any other animal in this book, the Horse has just one huge toe on each foot. This toe leaves an oval print with a distinctive 'frog' (V-shaped mark) at its base. If a Horse is shod, the horseshoe shows up clearly as a firm wall at the outside of the print. Not all horses are shod, so do not expect to see this outer wall on every horse print. A typical, unhurried horse trail shows an alternating walking pattern, with the hind prints registered on or behind the slightly larger fore prints. Horses are capable of a range of speeds—up to a full gallop—but most recreational horseback riders take a more leisurely outlook on life, preferring to walk their horses.

Similar Species: Mules (rarely shod) have smaller tracks.

Black Bear

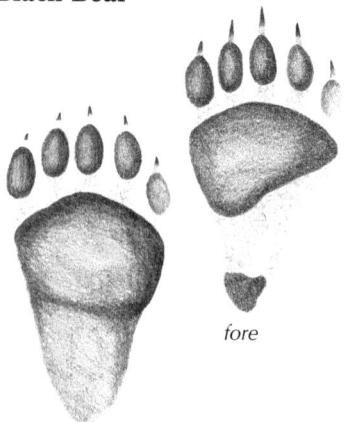

fore

hind

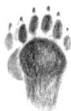

Fore Print
Length: 4–6.3 in (10–16 cm)
Width: 3.8–5.5 in (9.5–14 cm)

Hind Print
Length: 6–7 in (15–18 cm)
Width: 3.5–5.5 in (9–14 cm)

Straddle
9–15 in (23–38 cm)

Stride
Walking: 17–23 in (43–58 cm)

Size (male>female)
Height: 3–3.5 ft (90–110 cm)
Length: 5–6 ft (1.5–1.8 m)

Weight
200–600 lb (90–270 kg)

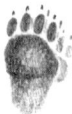

walking

BLACK BEAR
Ursus americanus

   The Black Bear has a much reduced and scattered
range in remote forested areas of this region. Finding
fresh bear tracks can be a thrill, but take care—the bear
may be just ahead. Never underestimate the potential
power of a surprised bear!

   Black Bear prints somewhat resemble small human
prints, but they are wider and show claw marks. The
small inner toe rarely registers. The forefoot's small heel
pad often registers, and the hind print shows a big heel.
The bear's slow walk results in a slightly pigeon-toed
double register with the hind print on the fore print.
More frequently, at a faster pace, the hind foot oversteps
the forefoot. When a bear runs, the two hind feet register
in front of the forefeet in an extended cluster. Along well-
worn bear paths, look for 'digs' (patches of dug-up earth)
and 'bear trees' whose scratched bark shows that this
bear climbs.

Similar Species: No other animal in this region leaves
similar tracks.

Domestic Dog

fore

hind

Fore Print (hind print is smaller)
Length: 1–5.5 in (2.5–14 cm)
Width: 1–5 in (2.5–13 cm)
Straddle
1.5–8 in (3.8–20 cm)
Stride
Walking: 3–32 in (7.5–80 cm)
Loping to Galloping: to 9 ft (2.7 m)
Size
Very variable
Weight
Very variable

walking

loping to galloping

DOMESTIC DOG
Canis familiaris

Dogs come in many shapes and sizes, from the tiny Chihuahua with its dainty feet to the robust and powerful Great Dane. Consequently, Domestic Dog tracks vary enormously. Dog ownership is high in many residential areas, and the popular pastime of dog walking can result in many dog tracks being left scattered about, especially if there is wet sand or mud.

The Domestic Dog's forefeet, which are much larger that the hind feet and support more of the weight, leave the clearest tracks. When a dog walks, the hind prints usually register ahead of or beside the fore prints. As the dog moves faster, it trots and then lopes before it gallops. In a trot or lope pattern the prints alternate fore-hind-fore-hind, whereas a gallop group shows (from back to front) fore-fore-hind-hind.

Similar Species: Keep in mind that dog prints are usually found close to human tracks or activity. Fox (pp. 30–33) prints may be confused with small dog prints. Coyote (p. 28) prints—also similar but usually more oval and splaying less—will be in a more direct trail.

Coyote

fore

hind

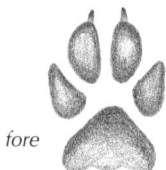

**Fore Print
(hind print is slightly smaller)**
Length: 2.4–3.2 in (6–8 cm)
Width: 1.6–2.4 in (4–6 cm)

Straddle
4–7 in (10–18 cm)

Stride
Walking: 8–16 in (20–40 cm)
Trotting: 17–23 in (43–58 cm)
Galloping/Leaping:
    2.5–10 ft (0.8 m–3 m)

Size (female is slightly smaller)
Height: 23–26 in (58–65 cm)
Length: 32–40 in (80–100 cm)

Weight
20–50 lb (9–23 kg)

*walking or
trotting*

*gallop
group*

COYOTE
(Brush Wolf, Prairie Wolf)
Canis latrans

This widespread, adaptable canine prefers open grass-lands or woodlands. On its own, with a mate or in a family pack, it hunts rodents and larger prey. If you find a coyote den—usually a wide-mouthed tunnel leading into a nesting chamber—do not bother the family or the female will have to move her pups to a safer location.

The hind print is slightly smaller than the oval fore print, and its less triangular heel pad rarely registers clearly. The claws of the two outer toes usually do not register. The Coyote typically walks or trots in an alternating pattern; the walk has a wider straddle, and the trotting trail is often very straight. When it gallops, the Coyote's hind feet fall in front of its forefeet; the faster it goes, the straighter the gallop group. The Coyote's tail hangs down, and it may leave a dragline in sand or loose dirt.

Similar Species: A Domestic Dog's (p. 26) less oval prints splay more, and its trail is erratic. Foot hairs blur Red Fox (p. 30) prints (usually smaller). Gray Fox (p. 32) prints are much smaller.

Red Fox

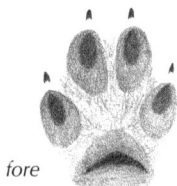

fore

hind

Fore Print
(hind print is slightly smaller)
Length: 2.1–3 in (5.3–7.5 cm)
Width: 1.6–2.3 in (4–5.8 cm)

Straddle
2–3.5 in (5–9 cm)

Stride
Trotting: 12–18 in (30–45 cm)
Side-trotting: 14–21 in (35–53 cm)

Size (vixen is slightly smaller)
Height: 14 in (35 cm)
Length: 22–25 in (55–65 cm)

Weight
7–15 lb (3.2–7 kg)

trotting   *side-trotting*

RED FOX
Vulpes vulpes

Very adaptable and intelligent, this beautiful and notoriously cunning fox is found throughout this region, in a variety of habitats from forests to open areas.

Abundant foot hair allows just parts of the toes and heel pads to register, with no fine detail. The horizontal or slightly curved bar across the fore heel pad is diagnostic. A trotting Red Fox leaves a distinctive straight alternating trail—the hind print direct registers on the wider fore print. When the fox side-trots, its print pairs show the hind print to one side of the fore print in typical canid fashion. This fox gallops like the Coyote (p. 28)— the faster the gallop, the straighter the gallop group.

Similar Species: Other canid prints lack the bar across the fore heel pad. Gray Fox (p. 32) prints are smaller. Domestic Dog (p. 26) prints can be of similar size, but they have a shorter stride and a less direct trail. Small Coyote prints are similar, but they have a wider straddle, and the toe marks are more bulbous.

Gray Fox

fore

hind

Fore Print
(hind print is slightly smaller)
Length: 1.3–2.1 in (3.3–5.3 cm)
Width: 1.1–1.5 in (2.8–3.8 cm)

Straddle
2–4 in (5–10 cm)

Stride
Walking/Trotting: 7–12 in (18–30 cm)

Size
Height: 14 in (35 cm)
Length: 21–30 in (53–75 cm)

Weight
7–15 lb (3.2–7 kg)

walking

GRAY FOX
Urocyon cinereoargenteus

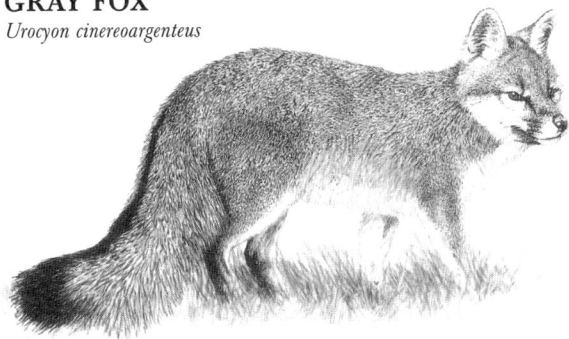

This small, shy fox is widespread, but it especially prefers woodlands and chaparral country. The Gray Fox is the only fox that climbs trees, which it does either for safety or to forage.

The forefoot registers better than the smaller hind foot, and the hind foot's long, semi-retractable claws do not always register. The heel pads are often unclear—they sometimes show up just as small, round dots. When it walks, this fox leaves a neat alternating track pattern. When it trots, its prints fall in pairs, with the fore print set diagonally behind the hind print. The Gray Fox's gallop group is like the Coyote's (p. 28).

Similar Species: The Red Fox (p. 30) has heel pads with a bar across them; its prints are generally larger and less clear (because of thick fur), its stride is longer, and its straddle is narrower. Coyote tracks are much larger. Domestic Cat (p. 38) and Bobcat (p. 36) prints lack claw marks and have larger, less symmetrical heel pads.

Mountain Lion

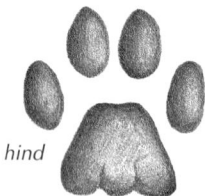

fore

hind

Fore Print
(hind print is slightly smaller)
Length: 3–4.5 in (7.5–11 cm)
Width: 3.3–4.8 in (8.5–12 cm)

Straddle
8–12 in (20–30 cm)

Stride
Walking: 13–32 in (33–80 cm)
Bounding: to 12 ft (3.7 m)

Size
Height: 25–32 in (65–80 cm)
Length: 3.5–5 ft (1.1–1.5 m)

Weight
70–200 lb (32–90 kg)

walking (fast)

MOUNTAIN LION
(Cougar, Puma, Panther)
Puma concolor

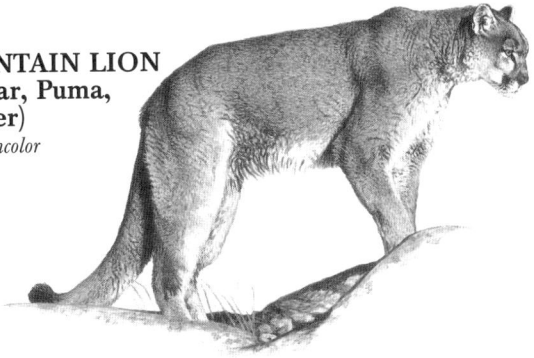

Shy, elusive and nocturnal, the Mountain Lion is spread widely but sparsely because of its need for a big home territory. Finding its tracks is usually the best that you can hope for. No longer common across much of its historic range, this cat and its tracks are infrequently reported in Mississippi and Louisiana.

Mountain Lion prints tend to be wider than they are long. The retractable claws never register. Thick foot fur enlarges the print in winter and may stop the two lobes on the front of the heel pad from registering clearly. In the walking gait, the hind print direct registers or double registers on the larger fore print. As the pace increases, the hind print tends to fall ahead of the fore print. The thick, long tail may occasionally leave a dragline that can blur print detail. A Mountain Lion seldom gallops, but it is capable of long bounds. Also look for partly buried scat and kills covered for later eating.

Similar Species: Bobcat (p. 36) prints may be confused with juvenile Mountain Lion prints. Canid (pp. 26–33) tracks show claw marks.

Bobcat

fore

hind

Fore Print
(hind print is slightly smaller)
Length: 1.8–2.5 in (4.5–6.5 cm)
Width: 1.8–2.5 in (4.5–6.5 cm)

Straddle
4–7 in (10–18 cm)

Stride
Walking: 8–16 in (20–40 cm)
Running: 4–8 ft (1.2–2.4 m)

Size (female is slightly smaller)
Height: 20–22 in (50–55 cm)
Length: 25–30 in (65–75 cm)

Weight
15–35 lb (7–16 kg)

walking

*ambling
to loping*

BOBCAT (Wildcat)
Lynx rufus

The widely distributed Bobcat, a stealthy and usually nocturnal hunter, is seldom seen. Very adaptable, it can leave tracks anywhere from wild mountainsides to chaparral and even in residential areas.

A walking Bobcat's hind feet usually register directly on its larger fore prints. As the Bobcat picks up speed, its trail becomes an ambling pattern of paired prints, the hind leading the fore. At even greater speeds, it leaves four-print groups in a lope pattern. The fore prints especially show asymmetry. The front part of the heel pad has two lobes and the rear part has three. The Bobcat's feet occasionally leave draglines.

Similar Species: In western Louisiana, the Ocelot (*Felis pardalis*) makes slightly larger and wider prints. A large Domestic Cat (p. 38) will make similar prints, but with a shorter stride and a narrower straddle. Fox (pp. 30–33), Domestic Dog (p. 26) and Coyote (p. 28) prints are narrower than they are long and show claw marks, and the fronts of their footpads are once-lobed.

Domestic Cat

fore

hind

Fore Print
(hind print is slightly smaller)
Length: 1–1.6 in (2.5–4 cm)
Width: 1–1.8 in (2.5–4.5 cm)

Straddle
2.4–4.5 in (6–11 cm)

Stride
Walking: 5–8 in (13–20 cm)
Loping/Galloping:
    14–32 in (35–80 cm)

Size (male>female)
Height: 20–22 in (50–55 cm)
Length with tail: 30 in (75 cm)

Weight
6.5–13 lb (3–6 kg)

walking

loping to galloping

DOMESTIC CAT
(House Cat)
Felis catus

The tracks of the familiar and abundant Domestic Cat can show up almost any place where there are people. Abandoned cats may roam farther afield; these 'feral cats' lead a pretty wild and independent existence. Domestic Cats come in many shapes, sizes and colors.

As with all felines, a Domestic Cat's fore print and slightly smaller hind print both show four toe pads. Its retractable claws, kept clean and sharp for catching prey, do not register. Cat prints usually show a slight asymmetry, with one toe leading the others. A Domestic Cat makes a neat alternating walking track pattern, usually in direct register, as one would expect from this animal's fastidious nature. When a cat picks up speed, it leaves clusters of four prints, the hind feet registering in front of the forefeet.

Similar Species: A small Bobcat (p. 36) may leave tracks similar to a very large Domestic Cat's. Domestic Dog (p. 26) and fox (pp. 30–33) prints show claw marks.

Raccoon

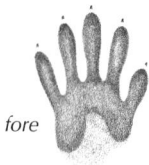

fore

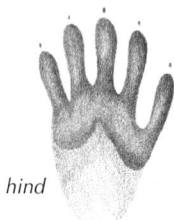

hind

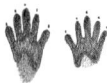

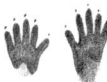

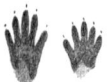

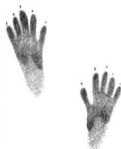

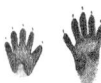

Fore Print
Length: 2–3 in (5–7.5 cm)
Width: 1.8–2.5 in (4.5–6.5 cm)
Hind Print
Length: 2.4–3.8 in (6–9.5 cm)
Width: 2–2.5 in (5–6.5 cm)
Straddle
3.3–6 in (8.5–15 cm)
Stride
Walking: 8–18 in (20–45 cm)
Bounding: 15–25 in (38–65 cm)
Size (female is slightly smaller)
Length: 24–37 in (60–95 cm)
Weight
11–35 lb (5–16 kg)

walking

*bounding
group*

RACCOON
Procyon lotor

The inquisitive Raccoon, which is common throughout this region, is adored by some people for its distinctive face mask, yet disliked for its boundless curiosity—often demonstrated with residential garbage cans. A good place to look for its tracks is near water at low elevations. The Raccoon likes to rest in trees. It usually dens up for the colder months.

The Raccoon's unusual print, showing five well-formed toes, looks like a human handprint; its small claws make dots. Its highly dexterous forefeet rarely leave heel prints, but its hind prints, which are generally much clearer, do show heels. The Raccoon's peculiar walking track pattern shows the left fore print next to the right hind print (or just in front) and vice versa. In mud, a Raccoon may leave a direct-registering pattern instead. The Raccoon occasionally bounds, leaving clusters with the hind prints in front of the fore prints.

Similar Species: In wet mud or loose dirt, unclear tracks by the Opossum (p. 42), River Otter (p. 46) or Woodchuck (p. 64) may look similar.

Opossum

fore

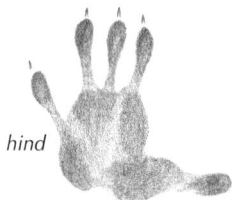

hind

Fore Print
Length: 2–2.3 in (5–5.8 cm)
Width: 2–2.3 in (5–5.8 cm)

Hind Print
Length: 2.5–3 in (6.5–7.5 cm)
Width: 2–3 in (5–7.5 cm)

Straddle
4–5 in (10–13 cm)

Stride
5–11 in (13–28 cm)

Size
Length: 2–2.5 ft (60–75 cm)

Weight
9–13 lb (4–6 kg)

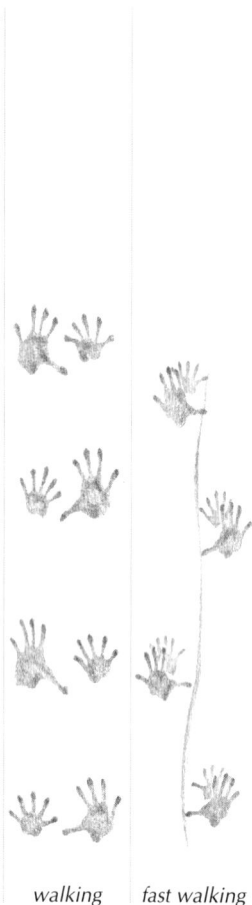

walking   fast walking

OPOSSUM
Didelphis virginiana

This slow-moving, nocturnal marsupial is found throughout the region. It occupies many habitats and is quite tolerant of residential areas, but it prefers open woodland or brushland around waterbodies. Look for Opossum tracks in mud near the water or near roadkill (which Opossums like to eat, though many of them suffer the same fate as the carrion that they dine on).

The Opossum has two walking habits: the common alternating pattern, with the hind prints registering on the fore prints, and a Raccoon-like (p. 40) paired-print pattern, with each hind print next to the opposing fore print. The very distinctive, long, inward-pointing thumb of the hind foot does not make a claw mark. The Opossum is an excellent climber, and its long, prehensile tail helps it maneuver amidst branches. However, contrary to popular myth, the Opossum does not regularly hang upside-down in trees by its tail.

Similar Species: Prints in which the distinctive thumbs do not show may be mistaken for a Raccoon's.

Nine-banded Armadillo

fore

hind

Fore Print
Length: 1.5–1.8 in (3.8–4.5 cm)
Width: 1.4–1.7 in (3.5–4.3 cm)

Hind Print
Length: 2–2.5 in (5–6.5 cm)
Width: 1.5–1.8 in (3.8–4.5 cm)

Straddle
2–3 in (5–7.5 cm)

Stride
3 in (7.5 cm)

Size
Length with tail: 24–32 in (60–80 cm)

Weight
8–17 lb (3.6–7.5 kg)

walking

NINE-BANDED ARMADILLO

Dasypus novemcinctus

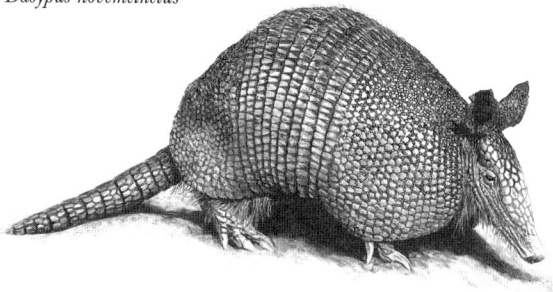

This comical, well-armored character constantly snuffles around through the dirt in a pig-like manner. If a Nine-banded Armadillo is preoccupied with its affairs, you can get surprisingly close, but hassle it too much and it may kick you with its strong legs. It can run quite quickly when it needs to, or it may curl up to protect its vulnerable underside.

Armadillo tracks, often in an alternating walking pattern, are usually most plentiful near their dens and burrows—look in sandy, loose soils that make for easy digging and foraging. A clear fore print shows four clawed toes, whereas a good hind print shows five. An armadillo's dragging tail or bony armor may wipe out print detail, and the armor may leave imprints during hot-weather mud baths. This animal can swim rivers or walk across on the bottom.

Similar Species: Clear armadillo prints are very bird-like. Unclear tracks in sand or dust may be mistaken for White-tailed Deer (p. 18) or Feral Pig (p. 20) prints.

River Otter

fore

hind

Fore Print
Length: 2.5–3.5 in (6.5–9 cm)
Width: 2–3 in (5–7.5 cm)

Hind Print
Length: 3–4 in (7.5–10 cm)
Width: 2.3–3.3 in (5.8–8.5 cm)

Straddle
4–9 in (10–23 cm)

Stride
Loping: 12–27 in (30–70 cm)

Size
(female is two-thirds the size of male)
Length with tail: 3–4.3 ft (90–130 cm)

Weight
10–25 lb (4.5–11 kg)

loping (fast)

RIVER OTTER
Lontra canadensis

No animal knows how to have more fun than a River Otter. If you are lucky enough to watch one at play, you will not soon forget the experience. Widespread and well-adapted for the aquatic environment, this otter lives near water. Expect to see a wealth of evidence of an otter's presence along the riverbanks in its home territory. The River Otter loves to slide down riverbanks, leaving troughs nearly 1 foot (30 cm) wide.

In soft mud, the webbing on the River Otter's five-toed feet, especially on the hind ones, may be evident. The hind foot's inner toe is set slightly apart from the rest. If the forefoot's metacarpal pad registers, it lengthens the print. Very variable, otter trails usually show a typical mustelid 2×2 loping, but with faster gaits the trails show groups of four and three prints. The thick, heavy tail often leaves a dragline.

Similar Species: Mink (p. 48) and Long-tailed Weasel (p. 50) prints are about half the size, with no conspicuous tail drag.

Mink

fore

hind

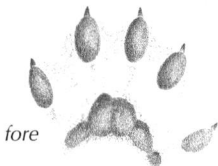

Fore and Hind Prints
Length: 1.3–2 in (3.3–5 cm)
Width: 1.3–1.8 in (3.3–4.5 cm)

Straddle
2.1–3.5 in (5.3–9 cm)

Stride
Walking/Loping: 8–36 in (20–90 cm)

Size (male>female)
Length with tail: 19–28 in (48–70 cm)

Weight
1.5–3.5 lb (0.7–1.6 kg)

2×2 loping

MINK
Mustela vison

The lustrous Mink, which is widespread throughout the region, prefers watery habitats surrounded by brush or forest. At home as much on land as in water, this nocturnal hunter can be exciting to track. Like the River Otter (p. 46), the Mink sometimes slides down banks, carving out a trough up to 6 inches (15 cm) wide.

The Mink's fore print shows five (perhaps four) toes, with five loosely connected palm pads in an arc, but the hind print shows only four palm pads. The metacarpal pad of the forefoot rarely registers, but the furred heel of the hind foot may register, lengthening the hind print. The Mink prefers the typical mustelid 2×2 loping gait, which leaves consistently spaced, slightly angled double prints. Its diverse track patterns also include alternating walking, loping with three- and four-print groups (like the River Otter) and bounding (like a rabbit, p. 56).

Similar Species: The Long-tailed Weasel (p. 50) makes similar but generally smaller tracks. The River Otter's trail shows distinct tail draglines. Bobcat (p. 36) prints may resemble four-toed Mink prints, but without claw marks or lobed palm pads and not in a bounding pattern.

49

Long-tailed Weasel

Fore and Hind Prints
Length: 1.1–1.8 in (2.8–4.5 cm)
Width: 0.8–1 in (2–2.5 cm)

Straddle
1.8–2.8 in (4.5–7 cm)

Stride
Bounding: 9.5–43 in (24–110 cm)

Size (male>female)
Length with tail: 12–22 in (30–55 cm)

Weight
3–12 oz (85–340 g)

2×2 loping

LONG-TAILED WEASEL
Mustela frenata

The Long-tailed Weasel is an
active year-round hunter, with an avid
appetite for rodents. It is found throughout Mississippi
and Louisiana, but its numbers may be declining. Following this nimble creature's tracks can reveal much
about its activities. A weasel trail may lead you up a tree
or to the edge of water—weasels are accomplished swim-
mers. Or it might disappear into a rodent burrow if the
weasel has pursued its occupant underground.

A weasel's light weight and small, hairy feet result in
pad detail that is often unclear. Even with clear tracks,
the inner (fifth) toe rarely registers. The usual weasel
gait is a 2×2 lope, leaving a trail of paired prints. The
Long-tailed Weasel's typical 2×2 lope shows an irregular
stride—sometimes short and sometimes long—with no
consistent behavior. Like the Mink (p. 48), this weasel
may bound like a rabbit (p. 56).

Similar Species: No other weasels are likely in this
region. Mink tracks, though similar, are generally larger.

Striped Skunk

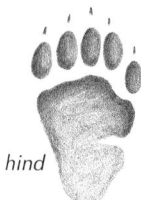

fore

hind

Fore Print
Length: 1.5–2.2 in (3.8–5.5 cm)
Width: 1–1.5 in (2.5–3.8 cm)

Hind Print
Length: 1.5–2.5 in (3.8–6.5 cm)
Width: 1–1.5 in (2.5–3.8 cm)

Straddle
2.8–4.5 in (7–11 cm)

Stride
Walking/Bounding:
2.5–8 in (6.5–20 cm)

Size
Length with tail:
20–32 in (50–80 cm)

Weight
6–14 lb (2.7–6.5 kg)

walking (fast)   bounding

STRIPED SKUNK
Mephitis mephitis

This striking skunk is notorious
for its vile smell, and the lingering odor is often the best
sign of its presence. Widespread throughout these two
states, it thrives in diverse habitats at lower elevations.
The Striped Skunk dens up in winter, coming out on
warmer days and in spring.

Both the forefeet and hind feet have five toes. The
long claws on the forefeet often register. The smooth
palm pads and small heel pads leave surprisingly small
prints. Skunks mostly walk—with such a potent smell for
their defense, they rarely need to run. Note that this
skunk's trail rarely shows any consistent pattern, but an
alternating walking pattern may be evident. The greater
a skunk's speed, the more the hind foot oversteps the
fore. If it runs, its trail consists of clumsy, closely set four-
print groups. It occasionally drags its feet.

Similar Species: The Eastern Spotted Skunk (p. 54)
makes smaller prints in a very random pattern. Mink
(p. 48) or Long-tailed Weasel (p. 50) tracks will be farther
apart than a skunk's. Skunk prints do not overlap.

Eastern Spotted Skunk

fore

hind

Fore Print
Length: 1–1.3 in (2.5–3.3 cm)
Width: 0.9–1.1 in (2.3–2.8 cm)

Hind Print
Length: 1.2–1.5 in (3–3.8 cm)
Width: 0.9–1.1 in (2.3–2.8 cm)

Straddle
2–3 in (5–7.5 cm)

Stride
Walking: 1.5–3 in (3.8–7.5 cm)
Bounding: 6–12 in (15–30 cm)

Size
Length: 13–25 in (33–65 cm)

Weight
0.6–2.2 lb (0.3–1 kg)

walking   *bounding*

EASTERN SPOTTED SKUNK
Spilogale putorius

   This beautifully marked skunk, smaller than its striped
cousin (p. 52), is found throughout most of the region,
but it may be declining in Louisiana. It enjoys diverse
habitats—such as scrubland, forests and farmland—but it
is a rare sight, because of its nocturnal habits, and because
it dens up in winter, coming out only on warmer nights.

   This skunk leaves a very haphazard trail as it forages
for food on the ground. Long claws on the forefeet often
register, and the palm and heel may leave defined pad
marks. Although this skunk rarely runs, when it does
so it may bound along, leaving groups of four prints,
hind in front of fore. It occasionally climbs trees, which
it does with ease. It sprays only when truly provoked,
so its powerful odor is less frequently detected than that
of the Striped Skunk.

Similar Species: The Striped Skunk has larger prints
and less scattered tracks with a shorter running stride
(or it jumps); it does not climb trees.

Eastern Cottontail

fore

hind

Fore Print
Length: 1–1.5 in (2.5–3.8 cm)
Width: 0.8–1.3 in (2–3.3 cm)

Hind Print
Length: 3–3.5 in (7.5–9 cm)
Width: 1–1.5 in (2.5–3.8 cm)

Straddle
4–5 in (10–13 cm)

Stride
Hopping: 0.6–3 ft (18–90 cm)

Size
Length: 12–17 in (30–43 cm)

Weight
1.3–3 lb (0.6–1.4 kg)

hopping

EASTERN COTTONTAIL
Sylvilagus floridanus

This abundant rabbit is found throughout Mississippi and Louisiana. Preferring brushy areas in grasslands and cultivated areas, it might be found in dense vegetation, hiding from predators such as the Bobcat (p. 36) and the Coyote (p. 28). Largely nocturnal, the Eastern Cottontail might be seen at dawn or dusk and on darker days.

As with other rabbits, this cottontail's most common track pattern is a triangular grouping of four prints, with the larger hind prints (which can appear pointed) in front of the fore prints (which may overlap). The hairiness of the toes will hide any pad detail. If you follow this rabbit's trail, you could be startled if it flies out from its 'form,' a depression in the ground in which it rests.

Similar Species: The Swamp Rabbit (*S. aquaticus*), common in swamps and bottomlands, makes similar but slightly larger prints. Squirrel (pp. 68–73) tracks show a similar pattern, but with the fore prints more consistently side by side.

Nutria

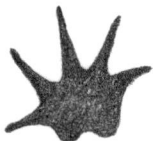

fore

hind

Fore Print
Length: to 3 in (7.5 cm)
Width: to 3 in (7.5 cm)
Hind Print
Length: 4.5–6 in (11–15 cm)
Width: to 3.5 in (9 cm)
Straddle
to 7 in (18 cm)
Stride
Walking: to 8 in (20 cm)
Size (male>female)
Length with tail: 25–55 in (65–140 cm)
Weight
5–25 lb (2.3–11 kg)

walking

NUTRIA
Myocastor coypus

Much larger than a Muskrat (p. 62), this rodent was introduced from South America by fur farmers. It has escaped into the wild, and there are now numerous colonies in the southern states. With a voracious appetite for rice and sugar cane, the Nutria can be quite destructive to agricultural activities.

The webbing on the Nutria's strong hind feet is often evident in its prints. Look for claw marks, too. The large hind print shows five toes, with the innermost toe set farther back than the others. The much smaller fore print also shows five toes. The large, round, hairless tail often leaves a dragline. A Nutria's trail may lead you to a den in a riverbank, possibly a former Muskrat residence. Nearby, you might also notice large mats of vegetation— a Nutria's feeding platform.

Similar Species: Beaver (p. 60) prints are similar but show webbing between all the toes of the hind foot. Muskrat prints are smaller and show no webbing.

Beaver

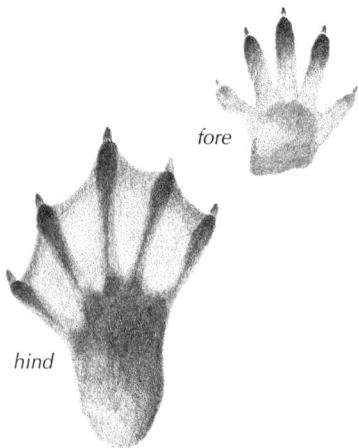

fore

hind

Fore Print
Length: 2.5–4 in (6.5–10 cm)
Width: 2–3.5 in (5–9 cm)
Hind Print
Length: 5–7 in (13–18 cm)
Width: 3.3–5.3 in (8.5–13 cm)
Straddle
6–11 in (15–28 cm)
Stride
Walking: 3–6.5 in (7.5–17 cm)
Size
Length with tail: 3–4 ft (90–120 cm)
Weight
28–75 lb (13–34 kg)

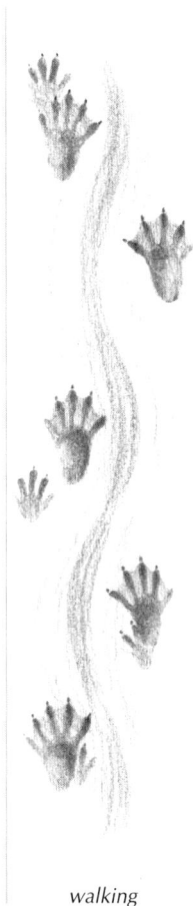

walking

BEAVER
Castor canadensis

Few animals leave as many signs of their presence as the Beaver, North America's largest rodent and a common sight around water. Look for the conspicuous dams and lodges—capable of changing the local landscape—and the stumps of felled trees. Check trunks gnawed clean of bark for marks of the Beaver's huge incisors. Scent mounds marked with castoreum, a strong-smelling yellowish fluid that Beavers produce, also indicate recent activity.

Check the large hind prints for signs of webbing and broad toenails. The nail of the fourth toe usually does not register, and it is rare for all five toes on each foot to do so. Irregular foot placement in the alternating walking gait may produce a direct register or a double register. The Beaver's thick, scaly tail may mar its tracks, as can the branches that it drags about for construction and food. Repeated path use results in well-worn trails.

Similar Species: The Beaver's many signs, including large hind prints, minimize confusion. Nutria (p. 58) hind prints lack webbing between the two fourth and fifth toes. Muskrat (p. 62) prints are much smaller.

Muskrat

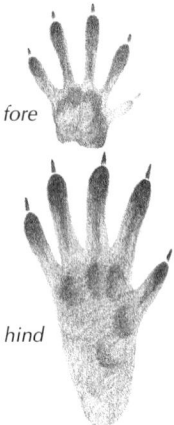

fore

hind

Fore Print
Length: 1.1–1.5 in (2.8–3.8 cm)
Width: 1.1–1.5 in (2.8–3.8 cm)

Hind Print
Length: 1.6–3.2 in (4–8 cm)
Width: 1.5–2.1 in (3.8–5.3 cm)

Straddle
3–5 in (7.5–13 cm)

Stride
Walking: 3–5 in (7.5–13 cm)
Running: to 1 ft (30 cm)

Size
Length with tail: 16–25 in (40–65 cm)

Weight
2–4 lb (0.9–1.8 kg)

walking

MUSKRAT

Ondatra zibethicus

Like the Beaver (p. 60), this rodent is found throughout the region, wherever there is water. Beavers are very tolerant of Muskrats and even allow them to live in parts of their lodges. Active all year, the Muskrat leaves plenty of signs. It digs extensive networks of burrows, often undermining riverbanks, so do not be surprised if you suddenly fall into a hidden hole! Also look for small lodges in the water and beds of vegetation on which the Muskrat rests, suns and feeds in summer.

The small fifth (innermost) toe of the forefoot rarely registers. Stiff hairs, which aid in swimming, may create a 'shelf' around the five well-formed toes of the hind print. The common walking pattern shows print pairs that alternate from side to side; the hind print is just behind the fore print or slightly overlaps it. In mud or sand, a Muskrat's feet can drag, and its tail usually leaves a sweeping dragline.

Similar Species: Few animals share this water-loving rodent's habits. The Beaver and the Nutria (p. 58) both make larger tracks, with webbing registering on the hind prints.

Woodchuck

fore

hind

Fore and Hind Prints
Length: 1.8–2.8 in (4.5–7 cm)
Width: 1–2 in (2.5–5 cm)

Straddle
3.3–6 in (8.5–15 cm)

Stride
Walking: 2–6 in (5–15 cm)
Bounding: 6–14 in (15–35 cm)

Size (male>female)
Length with tail:
    20–25 in (50–65 cm)

Weight
5.5–12 lb (2.5–5.5 kg)

walking   bounding

WOODCHUCK
(Whistle Pig, Groundhog, Marmot)
Marmota monax

This robust member of the squirrel family is occasionally seen in open woodlands and adjacent open areas in northern parts of the region. Always on the watch for predators, but not too troubled by humans, the Woodchuck never wanders far from its burrow. This marmot hibernates during winter but emerges in early spring; look for tracks in late spring in mud around the burrow entrances.

A Woodchuck's fore print shows four toes, three palm pads and two heel pads (not always evident). The hind print shows five toes, four palm pads and two poorly registering heel pads. The Woodchuck usually leaves an alternating walking pattern, with the hind print registered on the fore print. When a Woodchuck runs from danger, it makes groups of four prints, hind ahead of fore.

Similar Species: A small Raccoon's (p. 40) bounding track pattern will be similar, but it will show five-toed fore prints.

Eastern Chipmunk

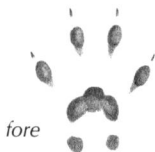

fore

hind

Fore Print
Length: 0.8–1 in (2–2.5 cm)
Width: 0.4–0.8 in (1–2 cm)

Hind Print
Length: 0.7–1.3 in (1.8–3.3 cm)
Width: 0.5–0.9 in (1.3–2.3 cm)

Straddle
2–3.2 in (5–8 cm)

Stride
Bounding: 7–15 in (18–38 cm)

Size
Length with tail: 7–10 in (18–25 cm)

Weight
2.5–5 oz (70–140 g)

bounding

EASTERN CHIPMUNK

Tamias striatus

Look for this delightful character in northern parts of the region. The Eastern Chipmunk is found in a variety of habitats, from the dense forest floor to open areas near buildings. You are more likely to see or hear this rodent, which is highly active during summer, than to notice its tracks. This large chipmunk is happiest on the ground, but it will gladly climb sturdy oak trees to harvest juicy, ripe acorns. It hibernates in winter, waking up from time to time to eat.

Chipmunks are so light that their tracks rarely show fine details. The forefeet each have four toes and the hind feet have five. Chipmunks run on their toes, so the two heel pads of the forefeet seldom register; the hind feet have no heel pads. Their erratic track patterns, like those of many of their cousins, show the hind feet registered in front of the forefeet. A chipmunk trail often leads to extensive burrows.

Similar Species: No other chipmunks inhabit this region. Squirrels (pp. 68–73) usually have larger prints and a wider straddle, and they are more likely to make midwinter tracks. Mouse (p. 84) tracks are smaller.

Eastern Gray Squirrel

fore

hind

Fore Print
Length: 1–1.8 in (2.5–4.5 cm)
Width: 1 in (2.5 cm)
Hind Print
Length: 2.3–3 in (5.8–7.5 cm)
Width: 1.1–1.5 in (2.8–3.8 cm)
Straddle
3.8–6 in (9.5–15 cm)
Stride
Bounding: 0.7–3 ft (22–90 cm)
Size
Length with tail: 17–20 in (43–50 cm)
Weight
14–25 oz (400–710 g)

bounding

EASTERN GRAY SQUIRREL
Sciurus carolinensis

This large, familiar squirrel can be a common sight in deciduous and mixed forests throughout this region, even in urban areas. Active all year, the Eastern Gray Squirrel can leave a wealth of evidence, especially in fall as it buries nuts for winter.

The Eastern Gray Squirrel leaves a typical squirrel track pattern when it runs or bounds. The hind prints fall slightly in front of the fore prints. A clear fore print shows four toes with sharp claws, four fused palm pads and two heel pads. The hind print shows five toes and four palm pads; if the full heel-length registers, it also shows two small heel pads.

Similar Species: Fox Squirrel (p. 70) prints are as big or larger. The Eastern Chipmunk (p. 66) and the Southern Flying Squirrel (p. 72) make smaller tracks in a similar pattern but with narrower straddles. Rabbits (p. 56) make longer track patterns, and their forefeet rarely register side by side when they run.

Fox Squirrel

fore

hind

Fore Print
Length: 1–1.9 in (2.5–4.5 cm)
Width: 1–1.7 in (2.5–4.3 cm)
Hind Print
Length: 2–3.3 in (5–7.5 cm)
Width: 1.5–1.9 in (3.8–4.8 cm)
Straddle
4–6 in (10–15 cm)
Stride
Bounding: 0.7–3 ft (22–90 cm)
Size
Length with tail: 18–28 in (45–70 cm)
Weight
1–2.4 lb (0.5–1.1 kg)

bounding

FOX SQUIRREL
Sciurus niger

This squirrel is much like the Eastern Gray Squirrel (p. 68), but it is larger and has a yellowish underside. It can be a common sight in deciduous forests with plenty of nut trees and in open areas or woodlands throughout the region. Its favorite feeding sites are indicated by piles of nutshells at tree bases. Active all year, the Fox Squirrel spends a lot of time foraging on the ground, often collecting nuts that it buried singly during the previous fall.

A clear fore print shows four toes with claws evident, four fused palm pads and two heel pads. The hind print shows five toes, four palm pads and sometimes a heel. When it runs or bounds, the Fox Squirrel makes a typical squirrel track, the hind prints slightly in front of the fore prints, with the prints in each pair roughly side by side.

Similar Species: The Eastern Gray Squirrel generally leaves smaller prints. Eastern Chipmunk (p. 66) and Southern Flying Squirrel (p. 72) tracks, which are in a similar pattern, are smaller and have narrower straddles. A rabbit (p. 56) makes a longer print pattern, and its fore prints rarely register side by side when it runs.

Southern Flying Squirrel

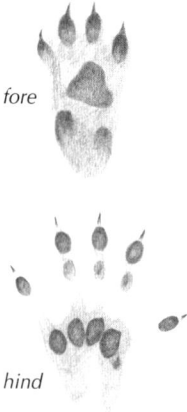

fore

hind

Fore Print
Length: 0.3–0.5 in (0.8–1.3 cm)
Width: 0.4 in (1 cm)
Hind Print
Length: 0.9–1.3 in (2.3–3.3 cm)
Width: 0.5 in (1.4 cm)
Straddle
2–2.5 in (5–6.5 cm)
Stride
Bounding: 7–22 in (18–55 cm)
Size
Length with tail: 8–10 in (20–25 cm)
Weight
1.5–3.2 oz (43–90 g)

bounding

SOUTHERN FLYING SQUIRREL
Glaucomys volans

This soft-furred brown acrobat is capable of long-distance gliding using the membranes that connect its forelegs and hind legs. Primarily nocturnal, the Southern Flying Squirrel is found in coniferous and mixed forests. In winter, up to 50 Southern Flying Squirrels can be found huddled together in a nest for warmth.

Flying squirrels spend so much of their time in trees and gliding in the air that they make few tracks. If one glides down to the ground, it can leave a distinctive four-print 'sitzmark,' but it often climbs down instead. New information from Mark Elbroch, a tracking expert, shows that the characteristic pattern for this squirrel is a bound where the front tracks register in front of the rear tracks, and the straddle is narrower than that of other squirrels.

Similar Species: Other squirrels (pp. 68–71) usually make larger prints and rarely leave sitzmarks, but with tracks in loose material it can be impossible to identify the squirrel species. Eastern Chipmunk (p. 66) tracks are smaller and have a narrower straddle.

Eastern Woodrat

fore

hind

Fore Print
Length: 0.6–0.8 in (1.5–2 cm)
Width: 0.4–0.5 in (1–1.3 cm)

Hind Print
Length: 1–1.5 in (2.5–3.8 cm)
Width: 0.6–0.8 in (1.5–2 cm)

Straddle
2.3–2.8 in (5.8–7 cm)

Stride
Walking: 1.8–3 in (4.5–7.5 cm)
Bounding: 5–8 in (13–20 cm)

Size
Length with tail:
    12–17 in (30–43 cm)

Weight
8–16 oz (230–450 g)

walking   *bounding*

EASTERN WOODRAT

Neotoma floridana

This nocturnal woodrat
thrives in rocky areas of Mississippi
and Louisiana. The trail of this woodrat
might lead you to a distinctive mass of a nest,
most often in a crevice, shrub or burrow. Although
it favors rocky areas, it feeds on foliage, seeds, ferns
and fungi and therefore requires nearby vegetation.

Four toes show on the fore print and five on the hind.
The short claws rarely register. A woodrat often walks
in an alternating fashion, with the hind print direct
registering on the fore print. This woodrat frequently
bounds as well, leaving a pattern of four prints, with
the larger hind print in front of the diagonally placed
fore print. The stride tends to be short relative to the
size of the prints.

Similar Species: Indistinct squirrel (pp. 68–73) prints
are similar, but they are usually larger. The Norway Rat
(p. 76) has similar prints, but it is usually found close to
human activity. Woodchuck (p. 64) prints are similar but
much larger.

Norway Rat

fore

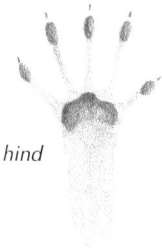

hind

Fore Print
Length: 0.7–0.8 in (1.8–2 cm)
Width: 0.5–0.7 in (1.3–1.8 cm)

Hind Print
Length: 1–1.3 in (2.5–3.3 cm)
Width: 0.8–1 in (2–2.5 cm)

Straddle
2–3 in (5–7.5 cm)

Stride
Walking: 1.5–3.5 in (3.8–9 cm)
Bounding: 9–20 in (23–50 cm)

Size
Length with tail: 13–19 in (33–48 cm)

Weight
7–18 oz (200–510 g)

walking

NORWAY RAT
(Brown Rat)
Rattus norvegicus

Active both day and night, this despised rat is widespread almost anywhere that humans have decided to build their homes. Not entirely dependent on people, it may live in the wild as well.

The fore print shows four toes, and the hind print shows five. When it bounds, this colonial rat leaves four-print groups, with the hind prints in front of the diagonally placed fore prints. Sometimes one of the hind feet direct registers on a fore print, creating a three-print group. This rat more commonly leaves an alternating walking pattern with the larger hind prints close to or overlapping the fore prints; the hind heel does not show. The tail often leaves a dragline in loose material. Rats live in groups, so you may find many trails together, often leading to their 5-inch (2-cm) wide burrows.

Similar Species: The Black Rat (*R. rattus*) and the Hispid Cotton Rat (p. 78), both also widespread, leave similar tracks. Mouse (p. 84) prints are much smaller. Chipmunk (p. 66) tracks are smaller, and the gaits differ.

Hispid Cotton Rat

fore

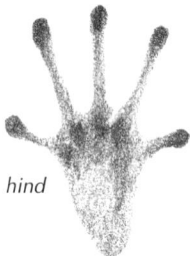

hind

Fore Print
Length: 0.5–0.7 in (1.3–1.8 cm)
Width: 0.5–0.7 in (1.3–1.8 cm)

Hind Print
Length: 0.6–1 in (1.5–2.5 cm)
Width: 0.6–0.8 in (1.5–2 cm)

Straddle
1.3–1.5 in (3.3–3.8 cm)

Stride
Walking: 1.3 in (3.3 cm)

Size
Length with tail: 8–14 in (20–35 cm)

Weight
2.8–7 oz (80–200 g)

walking

HISPID COTTON RAT
Sigmodon hispidus

   The Hispid Cotton Rat, which is found throughout these two states, makes itself unpopular by eating valuable crops. Though keen on devouring almost anything green, it prefers grassy fields. It stays close to home, and consequently the rat's little runways clearly mark its routes to favored feeding sites.

   The fore prints show four toes, but the larger hind prints usually show five. The heel of the hind foot will not always register, especially if the rat is moving fast. This medium-sized rodent leaves a typical walking track pattern in which the hind print registers on and slightly behind the fore print. Watch for its nests, which are balls of woven grass, and the small piles of cut grass it collects.

Similar Species: The Norway Rat (p. 76) and the Black Rat (*Rattus rattus*) also make similar tracks, but these two rats are usually found close to human activity.

Baird's Pocket Gopher

fore

hind

Fore Print
Length: 1 in (2.5 cm)
Width: 0.6 in (1.5 cm)

Hind Print
Length: 0.8–1 in (2–2.5 cm)
Width: 0.5 in (1.3 cm)

Straddle
1.5–2 in (3.8–5 cm)

Stride
Walking: 1.3–2 in (3.3–5 cm)

Size (male> female)
Length with tail: 7.5–8.75 in (19–22 cm)

Weight
0.25–0.5 oz (7–14 g)

walking

BAIRD'S POCKET GOPHER
Geomys breviceps

   This seldom-seen rodent of western Louisiana spends most of its time in burrows, venturing out only to move mud around and to find a mate. Because of its need to dig, the Baird's Pocket Gopher prefers soft, moist soils, and it especially enjoys pastureland.

   By far the best sign of pocket gopher activity is the muddy mounds that it creates when it burrows. As the gopher digs, it pushes the excess soil up out of the hole, creating a mound. Then, when it has finished digging, it pushes up a bit more soil to plug the entrance or 'cap' the burrow. Search around the mounds to find tracks. Each foot has five toes. Though the forefeet have long, well-developed claws for digging, the prints rarely show this much detail. Pocket gophers usually walk, leaving an alternating track pattern in which the hind prints fall on or slightly behind the fore prints.

Similar Species: Pocket gopher tracks are associated with their burrows. The Eastern Mole (p. 88) also leaves piles of pushed up soil (usually smaller), but it does not make distinct plugs. Moles leave 'ridges' as well.

Woodland Vole

fore

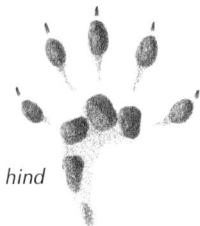

hind

Fore Print
Length: 0.5 in (1.3 cm)
Width: 0.5 in (1.3 cm)

Hind Print
Length: 0.6 in (1.5 cm)
Width: 0.5–0.8 in (1.3–2 cm)

Straddle
1.3–2 in (3.3–5 cm)

Stride
Walking/Trotting: 0.8 in (2 cm)
Bounding: 2–6 in (5–15 cm)

Size
Length with tail:
    4–5.5 in (10–14 cm)

Weight
0.8–1.3 oz (23–37 g)

walking

WOODLAND VOLE
(Pine Vole)
Microtus pinetorum

Distinguishing a vole track from those of the other small mammals in the region can be challenging. If you do spot a vole track, the only really likely candidate is the Woodland Vole. It lives in a variety of habitats throughout the two states.

When clear (which is seldom), vole fore prints show four toes, and hind prints show five. A vole's walk and trot both leave a paired alternating track pattern with a hind print occasionally direct registered on a fore print. Voles usually opt for a faster bounding; the resulting print pairs show the hind prints registered on the fore prints. This vole lopes quickly across open areas, creating a three-print track pattern. In summer, well-used vole paths appear as little runways in the grass. Look for distinctive piles of cut grass on runways near their ground nests. The bark at the bases of shrubs may show tiny teeth marks left by gnawing.

Similar Species: The only other vole to be found in the area is the Prairie Vole (*M. ochrogaster*), but it has an isolated population in southwestern Louisiana. Mouse (p. 84) bounding tracks show four-print groups.

Cotton Mouse

hind

fore

bounding group

Fore Print
Length: 0.3–0.4 in (0.8–1 cm)
Width: 0.3–0.4 in (0.8–1 cm)
Hind Print
Length: 0.3–0.6 in (0.8–1.5 cm)
Width: 0.3–0.4 in (0.8–1 cm)
Straddle
1.4–1.8 in (3.5–4.5 cm)
Stride
Bounding: to 12 in (30 cm)
Size
Length with tail:
    5.6–8 in (14–21 cm)
Weight
0.7–1.6 oz (20–45 g)

bounding

COTTON MOUSE
Peromyscus gossypinus

The nocturnal and seldom-seen Cotton Mouse is widespread throughout these two states. It prefers swampland, but it also frequents forests, rocky areas and even beaches. Its tracks may lead you up a tree, down a burrow or into a river—this mouse is a capable swimmer.

A clear Cotton Mouse fore print shows four toes, three palm pads and two heel pads. A hind print shows five toes (the fifth is often unclear) and three palm pads; the heel pads rarely register. Bounding tracks show the hind prints in front of the close-set fore prints. In wet mud or loose sand the prints may merge and appear as larger pairs, and tail drag will be evident.

Similar Species: Many less common species of mice have near-identical tracks. The House Mouse (*Mus musculus*) makes very similar tracks, but it associates more with humans. Voles (p. 82) tend to trot and have a much shorter bounding track pattern. The Eastern Chipmunk (p. 66) has a wider straddle. Shrews (p. 86) have a narrower straddle.

Southeastern Shrew

hind

fore

bounding group

Fore Print
Length: 0.2 in (0.5 cm)
Width: 0.2 in (0.5 cm)
Hind Print
Length: 0.6 in (1.5 cm)
Width: 0.3 in (0.8 cm)
Straddle
0.8–1.3 in (2–3.3 cm)
Stride
Bounding: 1.2–3 in (3–7.5 cm)
Size
Length with tail: 2.3–4.5 in (7–11 cm)
Weight
0.07–0.14 oz (2–4 g)

bounding

SOUTHEASTERN SHREW
(Bachman's Shrew)
Sorex longirostris

If you find shrew tracks in this region, the Southeastern Shrew is a likely candidate, even though several other species live in Mississippi and Louisiana. Unfortunately, recent studies indicate the Southeastern may be declining in Louisiana. This shrew prefers moist fields, marshes, bogs or woodlands, but it can also be found in higher and drier fields. Its rapid activity makes it difficult to observe closely.

In its energetic and unending quest for food, a shrew usually leaves a four-print bounding pattern, but it may slow to an alternating walking pattern. The individual prints in a group are often indistinct, but, in mud or moist sand, you can even count the five toes on each print. In loose sand or dirt, a shrew's tail may leave a dragline. A shrew's trail may disappear down a burrow.

Similar Species: The Southern Short-tailed Shrew (*Blarina carolinensis*) and the Least Shrew (*Cryptotis parva*) both make similar prints. The Pygmy Shrew (*S. hoyi*), also widespread, has slightly smaller prints. Mouse (p. 84) fore prints show four toes.

Eastern Mole

a molehill of the Eastern Mole

some molehills and ridges of the Eastern Mole

Size
Length with tail: 3.5–8.5 in (9–22 cm)
Weight
3–5 oz (85–140 g)

EASTERN MOLE

Scalopus aquaticus

This soft-furred resident of the underworld is the only mole that you will encounter in this region. It can leave a wealth of evidence that indicates its presence, usually in pastures or open woodlands, especially where the soil is light and moist and easy to burrow in.

Moles, which seldom emerge from their subterranean environment, create an extensive network of burrows through which they forage. These burrows are sometimes marked by ridges on the surface, though most of us are more familiar with the hills that form from the mole getting rid of excess soil from its burrows. Moles are frequently considered to be pests for messing up lawns with their earth works. When rain moistens the soil and bring worms to the surface, it can be entertaining to watch the earth twitch and rise up as the mole satisfies its voracious appetite.

Similar Species: No other moles live in this region. The mounds of the Baird's Pocket Gopher (p. 80) are larger and have distinct plugs that 'cap' the holes themselves.

BIRDS, AMPHIBIANS & REPTILES

A guide to the animal tracks of Mississippi and Lousiana is not complete without some consideration of the birds, amphibians and reptiles found in these states.

Several bird species have been chosen to represent the main types common to this region, but remember that individual bird species are not easily identified by track alone. Bird tracks are often abundant—the shores of lakes and streams are very reliable places to find them—especially in wet sand and mud, which can hold a clear print for a long time. The sheer number of tracks made by shorebirds and waterfowl can be astonishing. Though some bird species prefer to perch in trees or soar across the sky, it can be entertaining to follow the tracks of birds that spend a lot of time on the ground. They can spin around in circles and lead you in all directions. The trail may suddenly end as the bird takes flight, or it might terminate in a pile of feathers, the bird having fallen victim to a predator.

Many amphibians and turtles depend on moist environments, so look in the soft mud along the shores of lakes and ponds for their distinctive tracks. You may be able to distinguish frog tracks from toad tracks, because these two amphibians generally move differently, but it can be very difficult to identify the species. Reptiles thrive and outnumber the amphibians in drier environments, but they seldom leave good tracks, except in occasional mud or perhaps in sand. Snakes leave distinctive body prints.

Mallard

Print
Length: 2–2.5 in (5–6.5 cm)
Straddle
4 in (10 cm)
Stride
to 4 in (10 cm)
Size
23 in (58 cm)

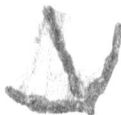

MALLARD
Anas platyrhynchos

female

male

This dabbling duck—the male a familiar sight with its striking green head—is common in open areas near lakes and ponds. Its webbed feet leave prints that can often be seen in abundance along the muddy shores of just about any waterbody, including those in urban parks.

The webbed foot of the Mallard has three long toes that all point forward. Though the toes register well, the webbing between the toes does not always show in the print. The inward-pointing feet give the Mallard a pigeon-toed appearance and perhaps account for its waddling gait, a characteristic for which ducks are known.

Similar Species: Many waterfowl, such as other ducks, as well as the Herring Gull (p. 94), leave similar prints. Exceptionally large prints were likely made by geese (various species) or by a swan (*Cygnus* spp.).

Herring Gull

Print
Length: 3.5 in (9 cm)
Straddle
4–6 in (10–15 cm)
Stride
4.5 in (11 cm)
Size
Length: 23–25 in (58–65 cm)

HERRING GULL
Larus argentatus

The Herring Gull, with its long wings and webbed toes, is a strong long-distance flier as well as an excellent swimmer. Increasingly common in the region, this bird is concentrated in great numbers near waterbodies and garbage dumps.

Gulls leave slightly asymmetrical tracks that show three toes. They have claws that register outside the webbing, and the claw marks are usually attached to the footprint. Most gulls have quite a swagger to their gait, and they leave a trail with the tracks turned strongly inward.

Similar Species: Gull species cannot be reliably identified by track alone, but smaller species have conspicuously smaller tracks. Mallard (p. 92) and other duck tracks are often difficult to distinguish from gull tracks. Geese (various species) make larger prints.

95

Great Blue Heron

Print
Length: to 6.5 in (17 cm)
Straddle
8 in (20 cm)
Stride
9 in (23 cm)
Size
4.2–4.5 ft (1.3–1.4 m)

GREAT BLUE HERON
Ardea herodias

The regal and graceful image of this large heron symbolizes the precious wetlands in which it patiently hunts for food. Usually still and statuesque as it waits for a meal to swim by, the Great Blue Heron will occasionally walk, perhaps to find a better hunting location. Look for its large, slender tracks along the banks or mudflats of waterbodies.

Not surprisingly, a bird that lives and hunts with such precision walks in a similar fashion, leaving straight tracks that fall in a nearly straight line. Look for the slender rear toe in the print.

Similar Species: Other herons and egrets (various species) have similar prints, varying in size relative to the size of the bird. Cranes (*Grus* spp.) occupy similar habitats and have similarly sized prints, but a crane's rear toes are smaller and do not register.

Common Snipe

Print
Length: 1.5 in (3.8 cm)
Straddle
to 1.8 in (4.5 cm)
Stride
to 1.3 in (3.3 cm)
Size
11–12 in (28–30 cm)

COMMON SNIPE
Gallinago gallinago

   This short-legged character is a resident of marshes and bogs, where its neat prints can often be seen in mud. Snipes are quite secretive when on the ground, and so you may be surprised if one suddenly flushes out from beneath your feet. If there is a Common Snipe in the air, you may hear an eerie whistle if it dives from the sky.

   The Common Snipe's neat prints show four toes, including a small rear toe that points inward. The bird's short legs and stocky body give it a very short stride.

Similar Species: Many shorebirds, including the Spotted Sandpiper (p. 100), leave similar tracks.

Spotted Sandpiper

Print
Length: 0.8–1.3 in (2–3.3 cm)
Straddle
to 1.5 in (3.8 cm)
Stride
Erratic
Size
7–8 in (18–20 cm)

SPOTTED SANDPIPER
Actitis macularia

The bobbing tail of the Spotted Sandpiper is a common sight on the shores of lakes, rivers and streams, but you will usually find just one of these territorial birds in any given location. Because of its excellent camouflage, likely the first sight that you will have of this bird will be when it flies away, its fluttering wings close to the surface of the water.

As a sandpiper teeters up and down on the shore, it leaves trails of three-toed prints that show a very small fourth toe that faces off to one side at an angle. Sandpiper tracks can have an erratic stride.

Similar Species: All sandpipers and plovers, including the common Killdeer (*Charadrius vociferus*), leave similar tracks, although there is much diversity in size. The Common Snipe (p. 98) makes similar but larger tracks.

Great Horned Owl

Strike
Width: to 3 ft (90 cm)
Size
22 in (55 cm)

GREAT HORNED OWL
Bubo virginianus

Often seen resting quietly in trees by day, this wide-ranging owl prefers to hunt at night. The mark that this accomplished hunter may leave in dust or sand can be quite a sight if it registers well. You might stumble across such a 'strike' and guess that the owl's target could have been a vole or mouse scurrying around on the ground. If you are a really lucky tracker, you might be following the trail of a small animal to find that it abruptly ends with this strike mark, where the animal has been seized.

The owl strikes with its talons, leaving scratch marks that are occasionally surrounded by the imprints of wings and tail feathers. These imprints result as the owl struggles to take off with possibly heavy prey. Not the most graceful of walkers, it prefers to fly away from the scene.

Similar Species: Other large birds, such as hawks, can also leave strike marks, but usually with less rounded, more distinct feather imprints.

American Crow

Print
Length: 2.5–3 in (6.5–7.5 cm)
Straddle
1.5–3 in (3.8–7.5 cm)
Stride
Walking: 4 in (10 cm)
Size
16 in (40 cm)

AMERICAN CROW
Corvus brachyrhyncos

The black silhouette of the American Crow can be a common sight in a variety of habitats. A crow will frequently come down to the ground and contentedly strut around with a confidence that hints at its intelligence. Its loud *caw* can be heard from quite a distance. Crows can be especially noisy when they are mobbing an owl or a hawk.

The American Crow typically leaves an alternating walking track pattern. Its prints show three sturdy toes pointing forward and one toe pointing backward. When a crow is in need of greater speed, perhaps for take-off, it bounds along, leaving irregular pairs of diagonally placed prints with a longer stride between each pair.

Similar Species: Other corvids, such as jays (various species), also spend a lot of time on the ground and make similar tracks that vary in size according to the size of the bird.

Northern Flicker

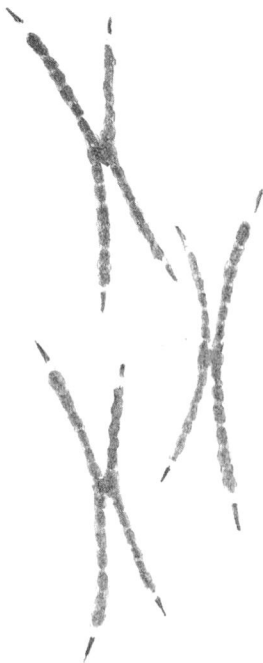

Print
Length: 1.8 in (4.5 cm)
Straddle
1–1.5 in (2.5–3.8 cm)
Stride
Hopping: 1.5–5 in (3.8–13 cm)
Size
5.5–6.5 in (14–17 cm)

NORTHERN FLICKER
Colaptes auratus

male

female

This attractive woodpecker, which can be seen throughout the region, is common in open woodlands and right into suburban areas. Unlike most woodpeckers, it spends some of its time feeding on the ground.

A clear flicker track shows a distinctive arrangement of two strong toes pointing forward and two pointing to the rear, with the outer toes slightly longer than the inner ones. The flicker's toes—along with short, strong legs that give the bird a short stride—are well suited for grasping tree trunks and limbs as this agile bird works its way along in search of insects.

Similar Species: Most other birds have very different tracks. Other woodpeckers would leave similar tracks, but very few come down to the ground as much as the obliging Northern Flicker does.

Dark-eyed Junco

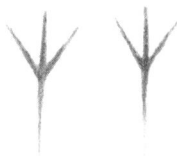

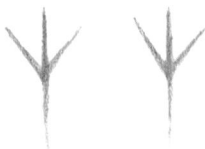

Print
Length: to 1.5 in (3.8 cm)
Straddle
1–1.5 in (2.5–3.8 cm)
Stride
Hopping: 1.5–5 in (3.8–13 cm)
Size
5.5–6.5 in (14–17 cm)

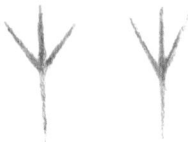

DARK-EYED JUNCO
Junco hyemalis

This common small bird typifies the many small hopping birds found in the region. Each foot has three forward-pointing toes and one longer toe at the rear. The best prints are left in moist, loose dirt, although sometimes the toe detail is lost. The footprints may show some dragging between the hops.

A good place to study this type of prints is near a birdfeeder. Watch the birds scurry around as they pick up fallen seeds, then have a look at the prints left behind. For example, the Dark-eyed Junco is attracted to seeds that chickadees (*Poecile* spp.) scatter as they forage for sunflower seeds in the birdfeeder. Also look for tracks under coniferous trees, where juncos feed on fallen seeds in winter.

Similar Species: The Northern Cardinal's (p. 110) tracks are very similar, as are those of many other small birds. Toe size may help with identification—larger birds make larger prints—as can the season. Sometimes junco tracks can be mistaken for mouse (p. 84) tracks, so follow the trail to see if it disappears down a hole or into thin air.

Northern Cardinal

Print
Length: to 1.5 in (3.8 cm)
Straddle
1–1.5 in (2.5–3.8 cm)
Stride
Hopping: 1.5–5 in (3.8–13 cm)
Size
9 in (23 cm)

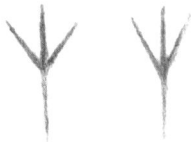

hopping

NORTHERN CARDINAL

Cardinalis cardinalis

female

male

The brilliant red plumage and small, black mask and chin of the male are a joy to see on this year-round resident. Like the Dark-eyed Junco (p. 108), the Northern Cardinal is a small hopping bird.

Each foot has three forward-pointing toes and one longer toe at the rear. The best prints are left in loose soil or sand. A good place to study these types of prints is near a birdfeeder. Watch the birds scurry around as they pick up fallen seeds, then have a look at the prints that they have left behind.

Similar Species: Juncos, finches (various species) and sparrows (various species) make similar tracks. The size of the toes may indicate what kind of bird you are tracking—larger birds have larger footprints. Not all birds are present all year, so keep in mind the season when tracking.

Frogs

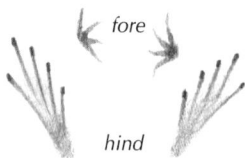

fore

hind

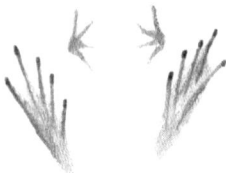

Straddle
to 3 in (7.5 cm)

hopping

FROGS

Bullfrog

There is a wide diversity of frogs in the region, all of which may leave tracks at some point. The Spring Peeper's (*Pseudacris crucifer*) 1.5-inch (3.8-cm) length and its preference for thick undergrowth and shrubs near the water make its tracks a rare sight. The larger gray tree-frogs (*Hyla chrysoscelis* and *H. versicolor*) spend most of their time in trees, coming down to breed and sing at night. The Green Frog (*Rana clamitans*) and the Pickerel Frog (*R. palustris*) both favor slow-moving, shallow water and swampy areas, and both are widespread. The beautiful and widespread Southern Leopard Frog (*R. spheno-cephala*), to 5 inches (13 cm) long, makes larger tracks. Unusually large tracks are surely from the robust Bullfrog (*R. catesbeiana*). Growing to 8 inches (20 cm) in length, it is North America's largest frog.

The best place to look for frog tracks is along the muddy fringes of waterbodies. A frog's hopping action results in its two small forefeet registering in front of its long-toed hind prints. Frog tracks vary greatly in size, depending on species and age. Toads (p. 114) usually walk, but they may also hop.

Toads

hind *fore*

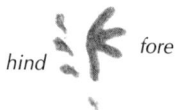

Straddle
to 2.5 in (6.5 cm)

walking

TOADS

Woodhouse's Toad

There are fewer toad species than frog species in the region. The toads most likely to be encountered, and the most widespread, are Woodhouse's Toad (*Bufo woodhousii*), which can be found in temporary pools and ditches, and the American Toad (*B. americanus*), which lives in many different moist habitats. In the far south, the tiny Oak Toad (*B. quercicus*) and the large, plump Southern Toad (*B. terrestris*) are also numerous. The Eastern Spadefoot (*Scaphiopus holbrookii*), found scattered in much of the region, lives in arid and semi-arid habitats. Toads in this region can be up to 5 inches (13 cm) long.

The best place to look for toad tracks is, undoubtedly, along the muddy fringes of waterbodies, but they can occasionally be found in drier areas— for example, as unclear trails in dusty patches of soil. In general, toads walk and frogs (p. 112) hop, but toads are pretty capable hoppers, too, especially when being hassled by overly enthusiastic naturalists. Toads leave rather abstract prints as they walk. The heels of the hind feet do not register. On less firm surfaces, the toes often leave draglines.

Salamanders & Newts

fore

hind

Straddle
to 4 in (10 cm)

walking

SALAMANDERS & NEWTS

Eastern Tiger Salamander

There are a wealth of salamanders and newts in the moist and wet areas of Mississippi and Louisiana. Among the more abundant and widespread of these long, slender, lizard-like amphibians is the Eastern Newt (*Notophthalmus viridescens*), which can grow to 5.5 inches (14 cm) in length. After a fresh rain, Eastern Newts emerging from ponds may leave small trails in the mud.

The Spotted Salamander (*Ambystoma maculatum*), which can grow to 10 inches (25 cm) long, lives throughout the region in mixed forests and some coniferous forests. The king of the salamander world is undoubtedly the magnificent Eastern Tiger Salamander (*A. tigrinum*), which can grow up to 13 inches (33 cm) long and comes in such a diversity of colors and patterns that it defies description. This heavy salamander leaves the best tracks, with a straddle of up to 4 inches (10 cm).

In general, a salamander fore print shows four toes, and the larger hind print shows five. However, print detail is often blurred by the animal's dragging belly or by the swinging of its thick tail across the tracks.

Lizards & Skinks

fore

hind

Straddle
to 3 in (7.5 cm)

walking

LIZARDS
& SKINKS

*Five-lined
Skink*

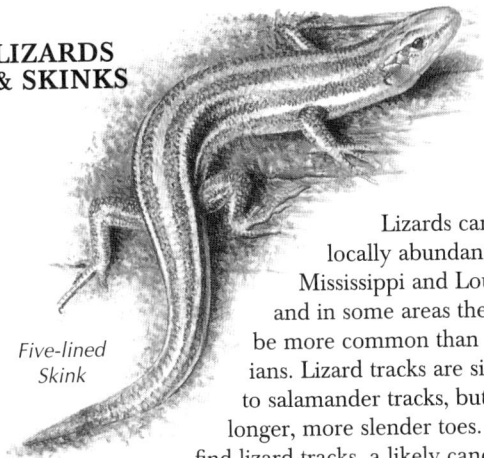

Lizards can be
locally abundant in
Mississippi and Louisiana,
and in some areas they can
be more common than amphibians. Lizard tracks are similar
to salamander tracks, but with
longer, more slender toes. If you
find lizard tracks, a likely candidate is
the Five-lined Skink (*Eumeces fasciatus*), which can grow
up to 8 inches (20 cm) long. It favors moist woodlands,
and it is widespread throughout the region. Another skink
that you might find is the Broadhead Skink (*E. laticeps*).

The Racerunner (*Cnemidophorus sexlineatus*) is also a
likely suspect. This lizard, which can grow to 11 inches
(28 cm) long, favors dry grasslands and well-drained
woodlands. The Fence Lizard (*Sceloporus undulatus*), which
grows up to 8 inches (20 cm) long, is common in open
woodlands and grasslands. The common Green Anole
(*Anolis carolinensis*), which is about the same size, favors
trees and shrubs, so it rarely leaves tracks.

These reptiles all move very quickly when the need
arises, their feet barely touching the ground as they dart
for cover. Consequently, their tracks can be hard to make
out clearly.

Turtles

fore

hind

Straddle
4–10 in (10–25 cm)

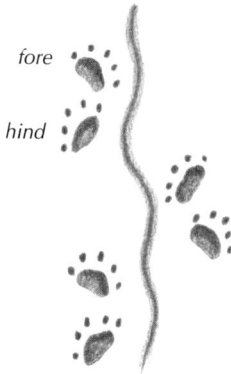

Snapping Turtle walking

fore

hind

typical turtle walking

TURTLES

*Eastern Box
Turtle*

Turtles,
those ancient
inhabitants of
the water world,
will happily slip into the murky depths to avoid detection,
but they do come out from time to time to feed or to bask
in the sunshine. Look for their distinctive tracks alongside
ponds, rivers and moist areas. Some turtles, such as the
huge Common Snapping Turtle (*Chelydra serpentina*),
rarely come onto land. One of the most widespread and
surprisingly terrestrial turtles is the Mud Turtle (*Kinosternon subrubrum*); it can reach almost 5 inches (13 cm) in
length. Slightly larger is the well-named Common Musk
Turtle (*Sternotherus odoratus*), which lives in shallow water
and produces a foul odor. The Slider (*Trachemys scripta*),
which reaches 12 inches (30 cm) in length, prefers slow-
moving rivers. The medium-sized Eastern Box Turtle
(*Terrapene carolina*) is common in moist forested areas.

With its large shell and short legs, a turtle leaves a
track pattern that has a straddle about half its body
length. Though longer-legged turtles can raise their
shells off the ground, short-legged species may let them
drag, which is shown in their tracks. The tail may leave
a straight dragline in the mud. On firmer surfaces, look
for distinct claw marks.

Snakes

SNAKES

Common Garter Snake

Many snake species inhabit the region, and they thrive in both arid and moist conditions. Because all snakes are long and slender, their tracks appear so similar that identification among the species is next to impossible. In fact, because a snake lacks feet and leaves a track that is just a gentle meander, it is very challenging even to establish in which direction a snake was moving.

The most widespread, frequently encountered snake is the harmless Common Garter Snake (*Thamnophis sirtalis*). Found throughout the region, often close to wet or moist areas, it can reach 4.3 feet (1.3 m) in length. Also widespread is the Redbelly Snake (*Storeria occipitomaculata*), which prefers hilly woodlands and can grow up to 16 inches (40 cm) long. Inhabiting grassy meadows and fields along forest edges, the Rough Green Snake (*Opheodrys aestivus*) can grow to 3.8 feet (1.2 m) in length. One of several rattlesnakes in the region is the Timber Rattlesnake (*Crotalus horridus*), which can grow up to 6.3 feet (1.9 m) long; it frequents a variety of habitats from marshlands to dry woodlands. Also common in a variety of habitats is the large and colorful Milk Snake (*Lampropeltis triangulum*).

TRACK PATTERNS & PRINTS

White-tailed Deer
p. 18

Feral Pig
p. 20

Horse
p. 22

Black Bear
p. 24

Domestic Dog
p. 26

Coyote
p. 28

Red Fox
p. 30

Gray Fox
p. 32

Mountain Lion
p. 34

Bobcat
p. 36

TRACK PATTERNS & PRINTS

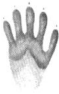

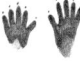

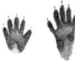

Domestic Cat
p. 38

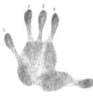

Raccoon
p. 40

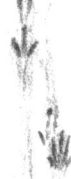

Opossum
p. 42

Nine-banded
Armadillo
p. 44

River Otter
p. 46

TRACK PATTERNS & PRINTS

Nutria
p. 58

Beaver
p. 60

Muskrat
p. 62

Woodchuck
p. 64

Eastern Chipmunk
p. 66

Eastern Gray
Squirrel
p. 68

Fox Squirrel
p. 70

Southern Flying
Squirrel
p. 72

Eastern Woodrat
p. 74

Norway Rat
p. 76

TRACK PATTERNS & PRINTS

Mallard
p. 92

Herring Gull
p. 94

Great Blue Heron
p. 96

Common Snipe
p. 98

131

TRACK PATTERNS & PRINTS

Spotted Sandpiper
p. 100

American Crow
p. 104

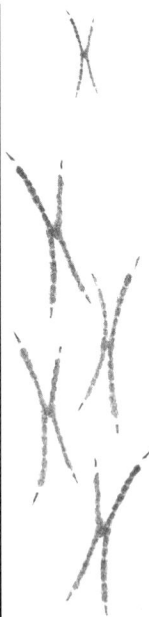

Northern Flicker
p. 106

Dark-eyed Junco
p. 108

Northern Cardinal
p. 110

Frogs
p. 112

Toads
p. 114

Salamanders &
Newts
p. 116

133

TRACK PATTERNS & PRINTS

Lizards & Skinks
p. 118

Typical Turtle
p. 120

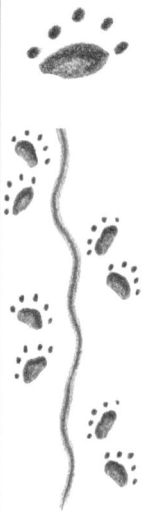

Common Snapping
Turtle
p. 120

Snakes
p. 122

HOOFED PRINTS

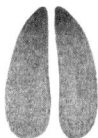

White-tailed
Deer

Horse

Feral Pig

```
inch   cm
0 ┬─ 0
  │
1 ┤
  │
2 ┴─ 5
```

HIND PRINTS

Cotton
Mouse

Southeastern
Shrew

Woodland
Vole

Hispid
Cotton Rat

```
inch   cm
0 ┬─ 0
  │
  ┤─ 1
  │
  ┤─ 2
1 ┤
  ┤─ 3
  │
  ┤─ 4
  │
2 ┴─ 5
```

Plains
Pocket Gopher

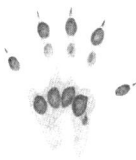

Southern
Flying Squirrel

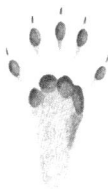

Eastern
Chipmunk

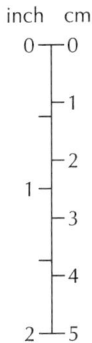

135

HIND PRINTS

Norway
Rat

Eastern
Woodrat

Woodchuck

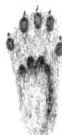

Muskrat

Fox
Squirrel

Eastern Gray
Squirrel

Eastern
Cottontail

inch cm
0 — 0
1
2 — 5

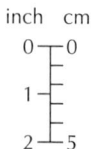

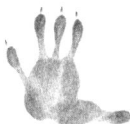

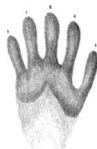

Nine-banded
Armadillo

Opossum

Raccoon

HIND PRINTS

inch cm
0 — 0
2
4 — 10

Nutria

Beaver

Black Bear

FORE PRINTS

Eastern
Spotted Skunk

Long-tailed
Weasel

Mink

Striped
Skunk

River
Otter

Gray Fox

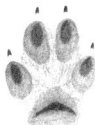

Red Fox

Coyote

Domestic Dog

Domestic
Cat

Bobcat

Mountain
Lion

inch   cm
0 — 0

1 —

2 — 5

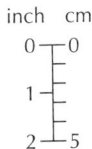

137

BIBLIOGRAPHY

Behler, J.L., and F.W. King. 1979. *Field Guide to North American Reptiles and Amphibians.* National Audubon Society. New York: Alfred A. Knopf.

Brown, R., J. Ferguson, M. Lawrence and D. Lees. 1987. *Tracks and Signs of the Birds of Britain and Europe: An Identification Guide.* London: Christopher Helm.

Burt, W.H. 1976. *A Field Guide to the Mammals.* Boston: Houghton Mifflin Company.

Farrand, J., Jr. 1995. *Familiar Animal Tracks of North America.* National Audubon Society Pocket Guide. New York: Alfred A. Knopf.

Forrest, L.R. 1988. *Field Guide to Tracking Animals in Snow.* Harrisburg: Stackpole Books.

Halfpenny, J. 1986. *A Field Guide to Mammal Tracking in North America.* Boulder: Johnson Publishing Company.

Headstrom, R. 1971. *Identifying Animal Tracks.* Toronto: General Publishing Company.

Murie, O.J. 1974. *A Field Guide to Animal Tracks.* The Peterson Field Guide Series. Boston: Houghton Mifflin Company.

Rezendes, P. 1992. *Tracking and the Art of Seeing: How to Read Animal Tracks and Signs.* Vermont: Camden House Publishing.

Stall, C. 1989. *Animal Tracks of the Rocky Mountains.* Seattle: The Mountaineers.

Stokes, D., and L. Stokes. 1986. *A Guide to Animal Tracking and Behaviour.* Toronto: Little, Brown and Company.

Wassink, J.L. 1993. *Mammals of the Central Rockies.* Missoula: Mountain Press Publishing Company.

Whitaker, J.O., Jr. 1996. *National Audubon Society Field Guide to North American Mammals.* New York: Alfred A. Knopf.

INDEX

Page numbers in **boldface** type refer to the primary (illustrated) treatments of animal species and their tracks.

ABOUT THE AUTHORS

Tamara Eder, equipped from the age of six with a canoe, a dip net and a note pad, grew up with a fascination for nature and the diversity of life. She has a degree in environmental conservation sciences and has photographed and written about the biodiversity in Bermuda, the Galapagos Islands, the Amazon Basin, China, Tibet and Southeast Asia.

Ian Sheldon, an accomplished artist, naturalist and educator, has lived in South Africa, Singapore, Britain and Canada. Caught collecting caterpillars at the age of three, he has been exposed to the beauty and diversity of nature ever since. He was educated at Cambridge University and the University of Alberta. When he is not in the tropics working on conservation projects or immersing himself in our beautiful wilderness, he is sharing his love for nature. Ian enjoys communicating this passion through the visual arts and the written word.